D1129746

Oceans and Seas

Arctic Ocean

John F. Prevost

ABDO Publishing Company

visit us at
www.abdopub.com

Published by ABDO Publishing Company, 4940 Viking Drive, Edina, Minnesota 55435.

Printed in the United States.

Photo Credits: Corbis, Corel Photo Disc

Contributing Editors: Kate A. Conley, Kristin Van Cleaf, Kristianne E. Vieregger
Art Direction & Maps: Neil Klinepier

Library of Congress Cataloging-in-Publication Data

Prevost, John F.
Arctic Ocean / John F. Prevost.
p. cm. -- (Oceans and seas)
Includes bibliographical references and index.
Summary: Surveys the origin, geological borders, climate, water, plant and animal life, and economic and ecological aspects of the Arctic Ocean.
ISBN 1-57765-095-6
1. Arctic Ocean--Juvenile literature. [1. Arctic Ocean.] I. Title. II. Series: Prevost, John F. Oceans and seas.
GC401.P74 1999
551.46'8--dc21 98-12071
CIP
AC

Contents

The Arctic Ocean 4
The Beginning 6
Climate 8
Arctic Waters 10
Plants 12
Animals 14
Native Peoples 16
Exploration 18
Today's Arctic 20
Glossary 22
How Do You Say That? 23
Web Sites 23
Index 24

The Arctic Ocean

The Arctic Ocean is the world's smallest ocean. It lies at the "top" of the world. The Arctic Ocean contains several bodies of water. They include the Barents **Sea**, the East Siberian Sea, and Baffin Bay.

The Arctic Ocean surrounds the North Pole. It is in the center of the **Arctic Circle**. The Great Bear constellation can be seen above the North Pole. This constellation gave the Arctic region its name. That's because the word *arctic* comes from the Greek word *arktos*, which means bear.

Many different **ethnic** groups live around the Arctic Ocean. It is bordered by six countries. These countries are Norway, Greenland, Russia, Canada, Iceland, and the United States.

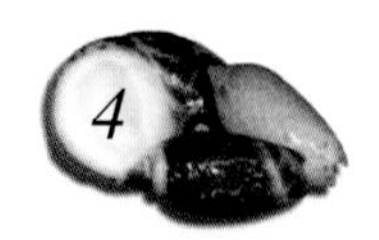

The natural resources in and around the Arctic Ocean are plentiful. However, pollution and overuse of the land have endangered some animal species. Arctic people are trying to preserve their ocean and land for the future.

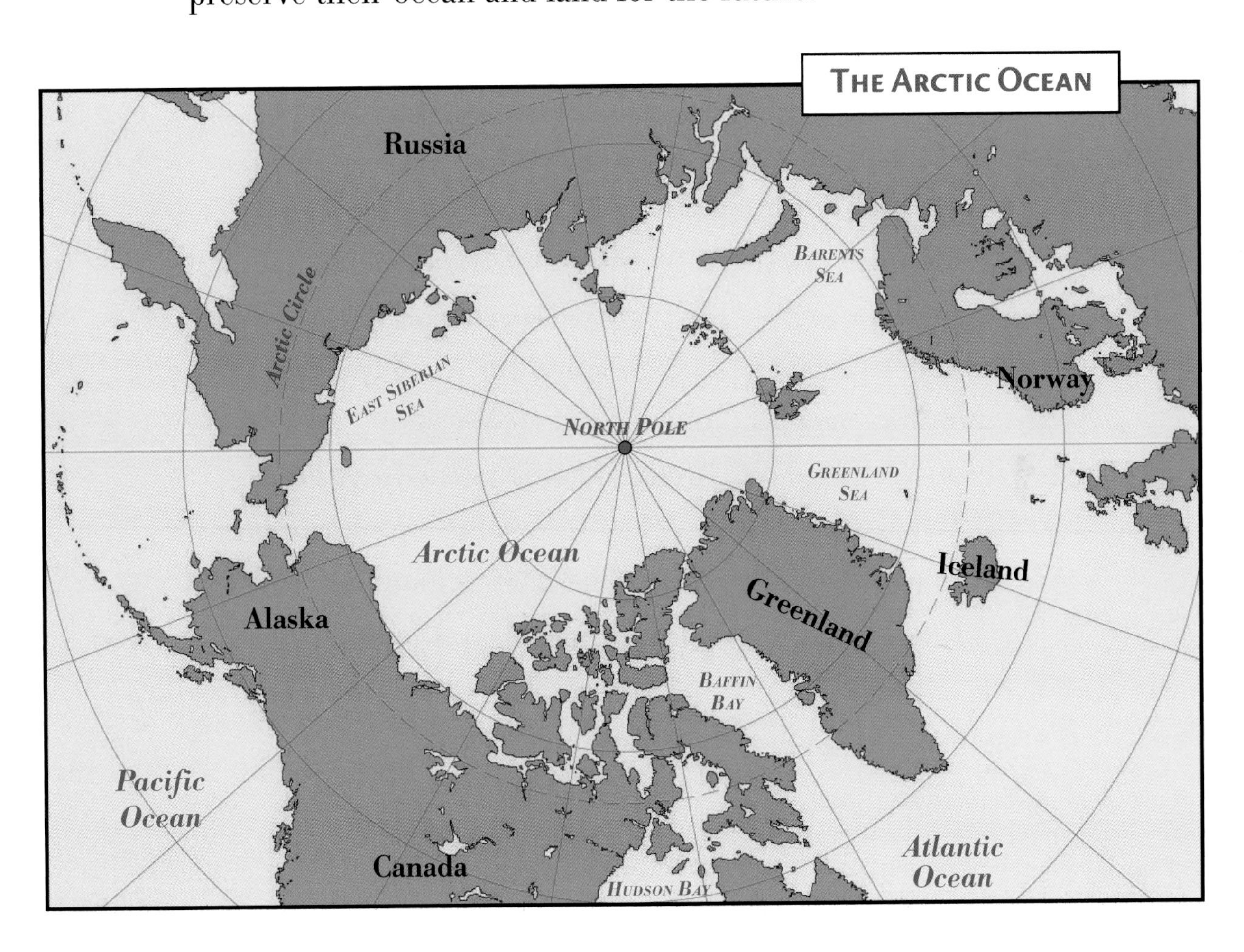

The Beginning

Scientists believe that millions of years ago Earth had only one ocean and one continent. The ocean was called Panthalassa. The continent was called Pangaea. Pangaea slowly split apart into today's seven continents. Panthalassa flowed between the continents and formed today's oceans.

After Pangaea split, **seafloor spreading** formed part of the Arctic Ocean's floor. The Arctic Ocean's floor has two huge basins. They are the Eurasia Basin and the Amerasia Basin. Together, they cover 5 million square miles (13 million sq km).

One-third of the Arctic Ocean's floor is **continental shelf**. The Arctic Ocean also has three underwater mountain ranges. The Lomonosov Ridge is the central mountain range. It is 1,100 miles (1,770 km) long. The deepest part of the ocean is in the Fram Basin. It is three miles (five km) deep. This area is a smaller basin found in the Eurasia Basin.

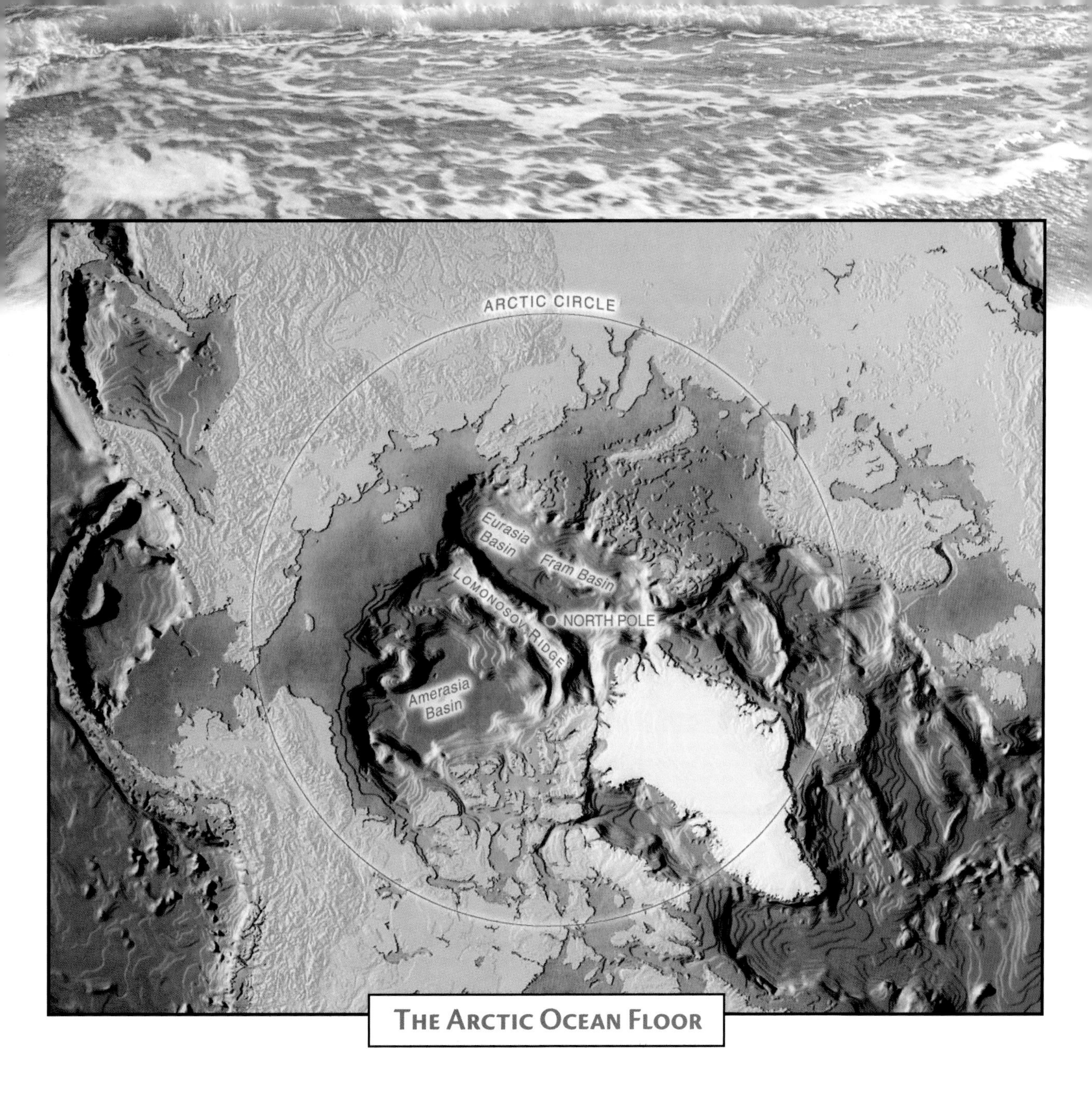

The Arctic Ocean Floor

Climate

The Arctic is the world's coldest ocean. Its water temperature changes very little with the seasons. During the winter, there is little sunlight. At this time, the water's average surface temperature is 28°F (-2°C).

By spring, ice covers almost all of the ocean. In summer, the sun never sets, which melts some of the ice. The surface water's average summer temperature is 29°F (-1.7°C).

The central part of the ocean, the area near the North Pole, is always covered with a huge, drifting **ice pack**. The ice averages 10 feet (3 m) thick, but may be many times thicker. During the winter, the ice pack extends to the surrounding land.

Whale hunting in the Arctic Ocean

Recently, the Arctic Ocean has experienced unusual changes in its climate. The ice in some areas is becoming thinner, and the air is warmer. This may be related to the **greenhouse effect**.

The Arctic Ocean near the North Pole

Arctic Waters

Ice covers much of the Arctic Ocean's water throughout most of the year. This ocean is almost completely icelocked from October to June. The ice stops shipping for the winter.

Icebergs in the Arctic Ocean can be dangerous. In the Arctic, icebergs form when chunks of ice break off glaciers in western Greenland and northeastern Canada. Every year, the Arctic produces about 50,000 icebergs. They can get in the way of boats.

During the spring and summer, the outer edge of the ice pack begins to melt and break apart.

Currents are streams of water flowing through the ocean. The Arctic Ocean receives most of its water from the North Atlantic Current. The East Greenland Current carries the majority of the water out of the Arctic Ocean. Several major rivers also bring water to the ocean.

Heavy currents keep the water moving. Sometimes, this stops icebergs from growing too large. Most of the tides in the Arctic Ocean rise no higher than one foot (30 cm).

Even large ships look quite small when next to an iceberg.

Plants

The cold temperatures and little sunlight prevent most plants from surviving in the Arctic Ocean. Plants need sunlight for **photosynthesis**. So the plants that do survive must grow close to the water's surface.

Currents move small, free-floating ocean plants called phytoplankton around in the water. When some of the **ice pack** breaks up in the summer, phytoplankton grow in the open water. Several ocean animals survive off their nutrients.

Ice algae also grow in the Arctic Ocean. These plants are found on the underside of thin ice, where sunlight can pass through. In the summer, a type of brown algae called kelp grows in the shallow waters. People harvest kelp for many reasons. Kelp can be eaten, and even made into candy!

The land around the Arctic Ocean is a treeless, grassy area called tundra.

Animals

People once believed that animals could not exist in the Arctic. In reality, many animals can survive the cold and ice. Fish, birds, whales, seals, and polar bears live in and around the Arctic Ocean. In fact, several of these species are found only in the Arctic region.

A beluga whale

Beluga whales are one of three types of whales that spend all their lives in the Arctic's waters. People sometimes call them "**sea** canaries." Like canaries, these whales make many different sounds. They like to spend a lot of time with each other, too.

Seals swim in the Arctic Ocean. They eat a lot of fish, as well as other ocean animals. Seals' toes are pointed backward. So instead of walking, they prefer to swim or slide around on the ice.

Polar bears are often seen hunting around the ocean. They hunt for seals and other animals. These dangerous animals can weigh as much as 1,000 pounds (454 kg)!

A polar bear

Native Peoples

Norway, Iceland, Greenland, Russia, Canada, and the United States surround the Arctic Ocean. Several different groups of native peoples live along the Arctic's coasts.

Native groups include the Alaskan Inuit and Aleut, the Canadian and Greenlandic Inuit, the Russian Nenets and Khants, and many others. Some people of European ancestry also live in the Arctic region.

Today, these native **cultures** struggle to keep their lands, resources, and identities. In Russia, some people want to use the land for its oil and other natural resources. But native peoples, such as the Khants, need this land for reindeer pastures. In Alaska, whale and seal **conservation** programs reduce hunting and the food available to Alaska Natives.

Many cultures living near the Arctic Ocean have been there for hundreds of years. They still use traditional methods of survival, while enjoying modern technologies, too.

A Greenlander teaches children how to make and use a traditional kayak.

Exploration

Explorers have been interested in the Arctic region for hundreds of years. The Greek explorer Pytheas sailed close to the **Arctic Circle** in the late 300s B.C. Many early explorers tried to reach the North Pole. Others searched for a northwest or northeast passage for ships to pass between the continents. Still others hunted whales and other animals for money.

Fridtjof Nansen

In A.D. 1893, a Norwegian man named Fridtjof Nansen set sail in search of the North Pole. He hoped that the currents would drift his boat, the *Fram*, across the North Pole. But the currents started bringing him too far south. So he decided to finish his journey by foot. He came close, but eventually had to turn back because of bad weather.

American Robert Edwin Peary also attempted to reach the North Pole. He claimed to have reached it in 1909, but he could not offer definite proof. However, he probably came within a few miles of it.

Robert Peary

In 1977, a Soviet nuclear-powered icebreaker broke through the Arctic **ice pack**. It was the first surface ship to reach the North Pole. In the 1990s, the U.S. Navy developed a submarine that could break through several feet of ice. This helped scientists further explore the ocean's floor and other characteristics.

Today's Arctic

Beautiful scenery, history, and **culture** make the Arctic Ocean's coasts great places for tourists. For example, adventure seekers travel the Arctic's icy waters with **kayaks** or other small boats.

In addition to tourism, the Arctic Ocean has many natural resources. However, the use of these resources has created problems. Hunting, fishing, and pollution have caused the near extinction of several species of the Arctic's animals. Today, many laws protect the Arctic Ocean and the wildlife that lives there.

Much of the Arctic Ocean and its surrounding area are still unexplored. Scientists are learning more about the area's resources and environmental history. They are improving the quality of life for humans, as well as animals in this region.

Snowmobiles are a useful form of transportation in some parts of the Arctic.

Glossary

Arctic Circle - an imaginary line around Earth's "top" that runs parallel to the equator at about 66° north latitude.

conservation - protection from loss or from being used up.

continental shelf - part of a continent that slopes gently away from the shoreline.

culture - the customs, arts, and tools of a nation or people at a certain time.

ethnic - a way to describe a group of people who have the same race, nationality, or culture.

greenhouse effect - the warming of Earth's surface, possibly caused by pollution.

ice pack - also called pack ice. A large layer of floating ice formed by pieces of ice that pressed together and froze into a single mass.

kayak - a traditional, one-person Inuit canoe that is entirely covered in seal skins. Today, kayaks are made of fiberglass.

photosynthesis - the process by which green plants use light energy, carbon dioxide, and water to make food and oxygen.

sea - a body of water that is smaller than an ocean and is almost completely surrounded by land.

seafloor spreading - a process that forms new seafloor.

How Do You Say That?

algae - AL-jee
Fridtjof Nansen - FRIH-chawf NAHNT-suhn
kayak - KI-ak
Pangaea - pan-JEE-uh
Panthalassa - pan-THA-luh-suh
photosynthesis - foh-toh-SIN-thuh-suhs
phytoplankton - fi-toh-PLANGK-tuhn
Pytheas - PITH-ee-uhs

Web Sites

Would you like to learn more about the Arctic Ocean? Please visit **www.abdopub.com** to find up-to-date Web site links about the Arctic Ocean's past, present, and future. These links are routinely monitored and updated to provide the most current information available.

Index

A
Amerasia Basin 6
animals 5, 12, 14, 15, 18, 20
Arctic Circle 4, 18

B
Baffin Bay 4
Barents Sea 4

C
Canada 4, 10, 16
climate 8, 9
currents 11, 12, 18

E
East Greenland Current 11
East Siberian Sea 4
Eurasia Basin 6

F
Fram Basin 6

G
Great Bear constellation 4
Greenland 4, 10, 16

I
ice pack 8, 12, 19
icebergs 10, 11
Iceland 4, 16

L
Lomonosov Ridge 6

N
Nansen, Fridtjof 18
native peoples 16
natural resources 5, 16, 20
Navy, U.S. 19
North Atlantic Current 11
North Pole 4, 8, 18, 19
Norway 4, 16, 18

O
ocean floor 6, 19

P
Pangaea 6
Panthalassa 6
Peary, Robert Edwin 19
plants 12
pollution 5, 20
Pytheas 18

R
Russia 4, 16

S
Soviet Union 19

T
tourism 20

U
United States 4, 16, 19

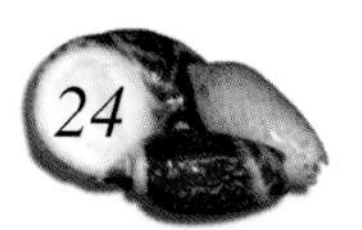